Keith Haddock

Iconografix

Iconografix
PO Box 446
Hudson, Wisconsin 54016 USA

Iconografix books are offered at a discount when sold in quantity for promotional use. Businesses or organizations seeking details should write to the Marketing Department, Iconografix, at the above address.

Library of Congress Control Number: 2011936671

ISBN-13: 978-1-58388-290-0
ISBN-10: 1-58388-290-1

11 12 13 14 15 16 6 5 4 3 2 1

Printed in China

On the cover: One of the world's largest shovels removes overburden in an Alberta, Canada, oilsands mine. The 4100C BOSS, the latest from P&H Global, employs AC electric motor drive and weighs 1,690 tons in operation. It is shown loading a Caterpillar 793D truck of 250 tons capacity.

TABLE OF CONTENTS

ACKNOWLEDGMENTS

I hope you enjoy the images I have assembled for this, my latest book on machines that move the earth. Unless otherwise noted the images are my own, taken during visits to earthmoving operations and surface mines in Canada, USA and the UK. Since I didn't have everything I needed, some of my professional and semi-professional photographer friends provided their own images to help fill gaps in my collection. A few other images were provided by manufacturers who responded instantly and professionally to my requests. To all the people listed below who helped make this book possible, I give my sincere thanks. Again I send a special "thank you" to my long-time friend and fellow author, Eric C. Orlemann, for the images and helpful advice he provided. Last but not least, another special "thank you" goes to my wife Barbara who did an expert job of editing my initial drafts, using experience from her school teaching days to considerably reduce the amount of editing required by the publisher!

Andreas Barner - Winkel, Switzerland
Brett Bedard & Audrie Roelf - John Deere
Eric Orlemann - ECO Communications
Gord King - SMS Equipment, Edmonton
Kelly Goossen - K-Tec Earthmovers Inc.
Sabrina Soares - Liebherr Mining Equipment
Shaun Herman - Strongco Equipment, Edmonton
Urs Peyer - Brunnen, Switzerland
Tim Twichell - Gowanda, New York
Tom McLaughlin - CNH Construction Equipment
Gavin Handley - Abigroup, Australia

ABOUT THE AUTHOR

The author, Keith Haddock, is a professional engineer. Born in Sheffield, England, he has enjoyed a lifelong involvement with earthmoving and heavy equipment. First in England and Scotland he was employed by Northern Strip Mining Ltd., and Tarmac Construction Ltd. and, starting in 1974, spent 24 years with Canadian surface mining company Luscar Ltd, latterly as Manager of Engineering. During that time he was directly involved with three new large surface coal mines with a production capability of six million tonnes of coal per year.

In 1986, Keith co-founded the Historical Construction Equipment Association (HCEA) based in Bowling Green, Ohio, which currently counts about 4,500 members. He has written or been involved with over 20 books on heavy equipment including Giant Earthmovers - an Illustrated History, Colossal Earthmovers, Extreme Mining Machines, Classic Caterpillar Crawlers, Bucyrus 125 Years, and Bucyrus Heavy Equipment. Keith has also been featured in numerous TV documentaries including Modern Marvels, Monster Machines, and Mega Excavators.

Currently Keith is principle of Park Communications, a consulting business specializing in providing information, research and freelance writing for earthmoving and surface mining publications. He currently writes for six magazines on a regular basis.

INTRODUCTION

Welcome again to the world of earthmoving machines. This book contains less historical information than my others. It is instead a full-color pictorial tribute to the marvelous machines working hard to enhance the high standard of living we enjoy today. Our transportation systems need roads, highways, railways, canals, docks, and airports. Our buildings require excavations for foundations and underground services. Our energy needs are served by power stations, coal mines, oil wells, pipelines and wind farms. Quarries and mines provide raw materials for bricks, cement, plastics, steel, aluminum, copper, and other metals we need for our computers, cars, home appliances, and other modern essentials. All of these commodities require a contribution from some type of earthmoving machine before they can be produced.

This book encompasses earthmoving machines of all sizes. While most of them can be seen at work on general construction sites, highways or industrial developments, the largest machines, usually found in surface mining operations on private property, are not often seen by the public. Visitors to these mines, unaccustomed to the gigantic proportions of the monster earthmoving machines, are often amazed by their size. My reply is, "What did you expect? A machine digging over 100 feet below, hoisting the material and dumping it over 400 feet away in one swoop, and doing it 24/7, has to be a giant."

Safety rules around earthmoving machines of all sizes are strict. Whether in vast mining operations or small construction sites, the same rules must be followed. Organization is crucial for efficiency on projects big or small because the problems encountered are the same. Haul roads must be kept smooth, equipment traffic routed to avoid congestion, water controlled and pumped, size of hauler and loader correctly matched, etc. Machine owners demand the lowest unit cost of material moved, so the most efficient machine size must be selected for each application; usually the larger the project, the larger the machine.

In choosing images for this book from the hundreds of available images suitable for publication, I experienced the usual difficulty in meeting the limited number allowed by the publisher. Thus my apologies to any manufacturers who think I should have shown more of their products; space simply did not permit. In the context of this book title, "Modern" means, with a few exceptions, machines that have been built in the last 15 years and, as far we know, are working today. Hopefully, readers familiar with the machines will appreciate those selected, and readers less familiar will gain an insight into the fascinating world of heavy equipment.

CHAPTER 1: BULLDOZERS

A new Caterpillar D10T helps to reclaim a surface coal mine in central Alberta following removal of the valuable resource. Its semi-U blade has an average pushing capacity of 24 cubic yards and, equipped with ripper as shown, boasts an operating weight of 146,400 pounds. The two exhaust stacks and the electronically-controlled fuel system ensure the 580-flywheel horsepower Caterpillar C27 engine meets EPA Tier 3 and EU Stage IIIa emission regulations.

Bulldozers are essential tools of any earthmoving operation. They perform a multitude of tasks on every project from initial access to final landscaping. The bulldozer is the ideal tool for haul road construction and maintenance, leveling off dump areas, assisting stuck vehicles, push-loading scrapers, clearing the area around large shovels to reduce tire damage on haul trucks, spreading gravel and other materials, and generally keeping the jobsite tidy. Although bulldozers primarily assist other earthmoving machines to do their job more efficiently, the larger sizes are sometimes assigned to production dozing in surface mines where overburden is pushed off the active pit into the mined-out area.

Over the decades, bulldozers have evolved from small tractors, wheeled or crawler, to the large efficient machines of today boasting over 1,000 horsepower. While most consist of a dozer blade mounted on the front of a crawler tractor, others run on four large rubber tires. Although not as robust for heavy dozing, wheel dozers boast superior mobility and higher travel speeds than their crawler counterparts.

This is Caterpillar's largest bulldozer to date. The massive D11T weighs 230,000 pounds when equipped with Universal blade and ripper, and its Caterpillar C32 engine puts out 850 flywheel horsepower. Caterpillar's unique raised crawler drive sprocket ("high drive"), traditional in all the company's large crawler tractors, permits a flexible undercarriage design for a smoother ride and convenient modular drive train component replacement. This D11T removes overburden in an Ohio surface coal mine.

Lots of activity in this central Indiana coal mine where a large fleet of Caterpillar D11R dozers rips and removes overburden. The D11R precedes the current D11T series tractors with the same weight and power. To increase efficiency still further in this hard material, a team of Drilltech and Ingersoll-Rand blast hole drills can be seen in the background.

At the small end of Caterpillar's extensive crawler tractor line is the D3K at 74 flywheel horsepower and 17,185 pounds operating weight. The machine pictured is equipped with an "AccuGrade" laser system for precise automatic grade control. The D3K, like Caterpillar's other smaller tractors (D4K, D5K and D6K), is hydrostatically driven, allowing independent stepless power on each track for power steering, or counter-rotation spin turns. *Andreas Barner*

Pushing a full blade of earth rated at eight cubic yards, this John Deere 850C dozer at 185 net horsepower was the largest built by the company up to the year 2000. The fully hydrostatic 850C features a load-sensing device that automatically adjusts speed and power to match load conditions. In 2000 Deere established a joint venture with Germany's Liebherr that resulted in larger hydrostatic-drive tractors. Deere has also updated its crawlers to the current J-series, which includes the 850J.

A rocky earthmoving job keeps this top-of-the-line John Deere dozer busy. The 324 net horsepower 1050C, with operating weight of 75,000 pounds, features hydrostatic drive and pilot-operated single lever control of speed, direction and steering. Deere's two largest crawler tractors are based on Liebherr designs from Germany and are powered by Liebherr engines. The latest versions are the 950J and 1050J which have slightly increased power ratings of 247 and 335 horsepower respectively. *Eric C. Orlemann*

The current line of Case crawler dozers is represented here by its flagship model 1650L at 144 horsepower and working weight of 36,000 pounds. Like most competitors in this size range, the L Series machines are hydrostatically driven, affording superior maneuverability with power to both tracks at all times and spin turns available when the tracks are counter-rotated. The inside-mounted blade can be angled from left to right or tilted forward or backward under power. *Case Construction Equipment*

All Liebherr crawler tractors and loaders have featured hydrostatic drive since the very first was launched in 1974. Although not the first in the industry, Liebherr persisted with this drive choice until perfection was reached. The latest Liebherr PR764, shown here with blade and ripper, is the world's largest hydrostatic drive bulldozer. It runs with a 422-horsepower Liebherr diesel engine and tips the scales at 116,150 pounds. *Liebherr*

New Holland crawler tractors are direct descendents from the former FiatAllis line, resulting from the formation of CNH Global in 1999. This group encompassed several well-known brand names (including Case), and is now marketed by New Holland Kobelco Construction Machinery S.p.A, one of the Fiat Group. The New Holland line is represented here by the 140-horsepower DC150 weighing in at 32,880 pounds. It is mechanically driven with automatic transmission.

Dressta equipment is built in Poland by Huta Stalowa Wola (HSW) and exported around the world. The TD-40E is the company's current largest crawler tractor, and carries a 515-horsepower Cummins QSK 19 diesel which meets EPA Tier 3 and EU Stage IIIa emission regulations. Transmission is a 3-speed power shift with electro-hydraulic control and torque converter. Operating weight with blade and ripper is 150,000 pounds. The TD-40E is a direct descendent of the former American-built International Dresser machines. *Marek Stankowski, HSW*

The 190-horsepower Komatsu 65PX-12 is a mid-size tractor from the company line that ranges from the smallest to the world's largest. The hydrostatic machine is controlled by a single joystick lever for steering, direction, and speed, and another joystick lever controls all blade movements. The 65PX model is a low ground pressure version with 36-inch shoes, while the 65EX version runs on standard 20.1-inch shoes.

Big crawler tractors are very good at ripping hard rock, extending the volume that can be moved without resorting to drilling and blasting. Here a Komatsu D475A shows what ripping is all about on a site development in Nevada. With an operating weight of 238,960 pounds when equipped with ripper and 'U' dozer blade, and 890 net horsepower, the D475A competes head to head with Caterpillar's largest bulldozer, the D11T. *Eric C. Orlemann*

Komatsu unveiled the first version of its huge D575A-2 bulldozer in 1991, and ever since this machine has held title to the world's largest production bulldozer. The current D575A-3 model's super power comes from a Komatsu diesel engine developing 1,150 net horsepower, and if scales could weigh this monster in operation, they would show a whopping 336,420 pounds. The blade measures 24 feet, 3 inches wide by 11 feet, 11 inches high at its center. Overall machine length is 38 feet, 5 inches.

Wheel dozers serve a multitude of tasks on large earthmoving projects and in surface mines. Although not robust enough for the tough jobs, their speed is an advantage over their crawler counterparts. Moving quickly between tasks, wheel dozers clear around large shovels to reduce truck tire damage, level dump areas, and maintain haul roads. The 854K is the flagship of Caterpillar's five-machine wheel dozer lineup. It is rated at 801 flywheel horsepower and 216,275 pounds operating weight. Its semi-U blade measures 20 feet, 8 inches wide.

The futuristic Caterpillar D7E is pioneering diesel-electric hybrid technology in crawler tractors. Its 235-net horsepower C9.3 diesel engine drives an AC electric generator which delivers electric current to a solid-state power inverter. Advanced electronics in the power inverter send AC current to the propulsion module to control the drive motors, and also provide DC current for tractor accessories. The electric drive motors deliver torque to the final drives, while steering functions through an updated differential steering system. *Urs Peyer*

CHAPTER 2: WHEEL & CRAWLER LOADERS

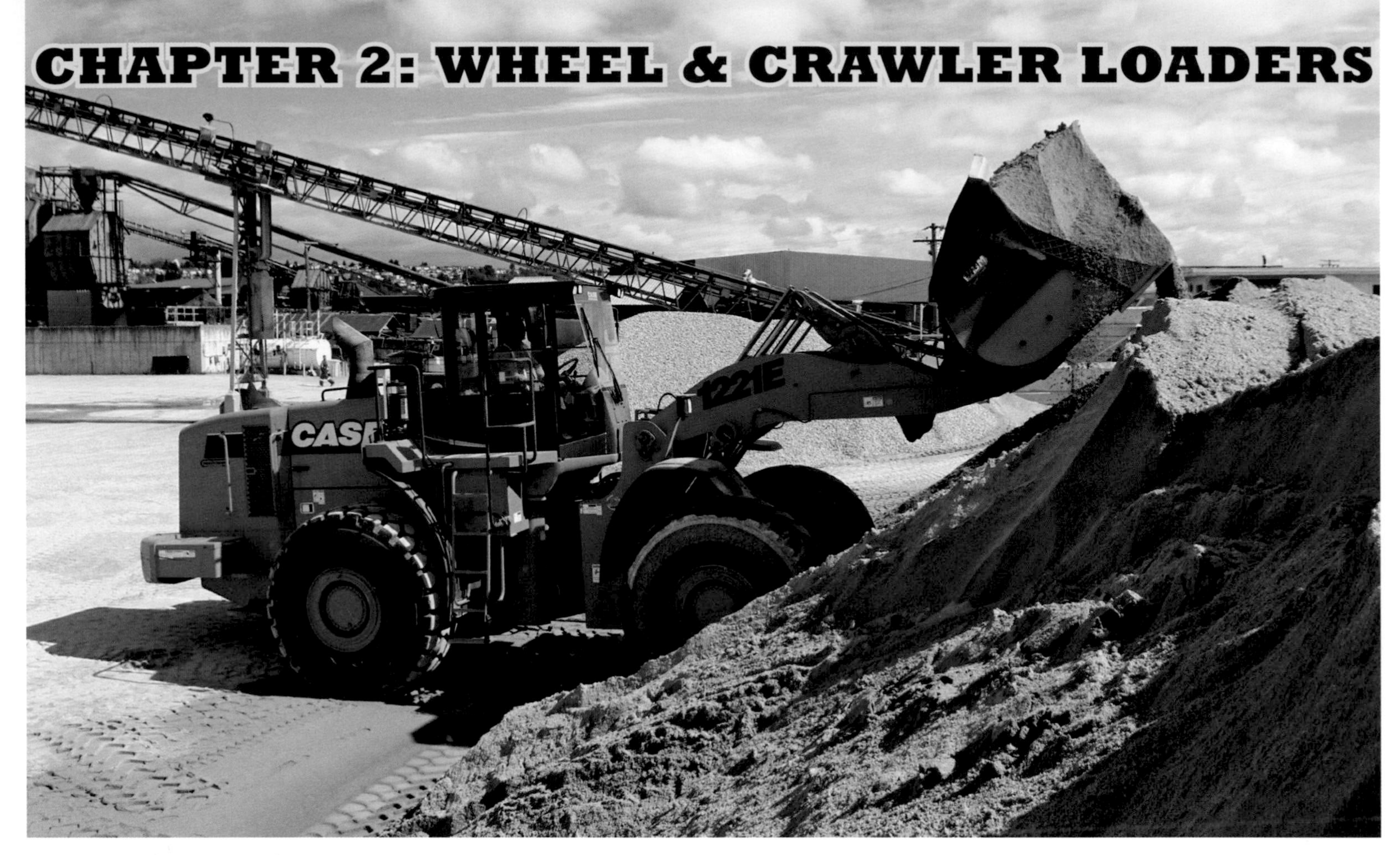

Heading up the wheel loader line offered by Case is the 1221E carrying a standard bucket of 7.6 heaped cubic yards. Power comes from a Cummins QSM 11, Tier III EPA-Certified engine putting out 320 net horsepower. A torque-sensing auto-shift transmission, with a choice of manual or fully-automatic control, delivers power to limited-slip differentials in each drive axle. The 1221E wheel loader is derived from a similar-sized Hyundai model, resulting from a joint venture between Case and Hyundai. *Case Construction Equipment*

Wheel loaders, and to a lesser extent crawler loaders, are some of the most popular machines seen today on construction sites, in quarries, gravel pits and surface mines. They originated in the 1920s when cable-controlled loader buckets were first mounted on farm tractors. Over the decades loaders have evolved into integrated, highly efficient production machines, designed from the ground up for digging. Modern machines boast articulated steering, advanced computer controlled transmissions, and fuel-efficient engines.

In many applications wheel loaders have largely taken over work done previously by crawler-mounted cable shovels. They are cheaper to buy and easier to transport. Their positive hydraulic bucket action and nimble articulated steering provide clear advantages over outmoded cable machines. Even in major quarries and surface mines where large electric rope shovels dominate, large wheel loaders are found to be valuable because of their adaptability and mobility. Mine owners have realized that even hard material can be moved profitably with wheel loaders if assisted by ripping or drilling and blasting.

Demonstrating just how tough wheel loaders are built today, this Caterpillar 990 wheel loader's job is to remove huge pre-cut pieces of blue limestone in a Belgian quarry. The hydraulics are powerful enough to raise the rear wheels off the ground. The 990 is at the upper end of Caterpillar's mid-size loader range with quarry bucket rated at 11.25 cubic yards. The latest 990H loader weighs 183,057 pounds and is supplied with a Caterpillar C27 engine of 627 net horsepower.

This Caterpillar 992G keeps busy loading rocky overburden into 150-ton haul trucks at an Ohio surface coal mine. The machine carries a 15.5 cubic yard mining bucket, runs with an 800-net horsepower Caterpillar engine, and tips the scales at 204,910 pounds. The 992G sports a revolutionary one-piece, box section, lift arm made from a steel casting. Improved operator visibility, increased torsional loading strength, and reduced overall weight are said to be its main advantages.

Pushing up in size is one of Caterpillar's latest wheel loaders, the 993K. Breaking tradition with its odd-number nomenclature for wheel loaders, the 993K in the 150-ton weight class uses a 950-net horsepower C32 engine. It handles a 17-yard bucket for rock or one up to 28 cubic yards for coal. An optional high-lift version increases its reach by about two feet to load high sided trucks. Unlike its smaller brother, the 992G, the 993K has double steel plate lift arms rather than a single casting. *Caterpillar*

The 994F is Caterpillar's flagship loader. Running on 53.5/85-57 tires, the huge loader weighs 427,209 pounds when working with a 25-yard standard bucket or 38-ton payload. It boasts a 16-cylinder Cat 3516B HD EUI turbocharged engine putting out 1,463 net horsepower, and requiring four stacks for exhaust. At surface mines, when a big shovel goes down, the 994F is able to temporarily take over excavating and loading duties, and can comfortably load a 250-ton truck in seven passes. *Eric C. Orlemann*

A crawler loader is more of an excavating tool than a wheel loader, offering greater stability and ability to work in soft underfoot conditions. Caterpillar offers a four-model range of hydrostatically-driven crawler loaders, headed by the 973D which handles a 4.2 cubic yard general purpose bucket. Operating weight is 61,857 pounds and a Cat C9 engine provides 263 net horsepower. Optional joystick controls allow the operator to control all machine travel and bucket movements with just two levers. *Andreas Barner*

John Deere is another manufacturer offering a full line of wheeled and crawler loaders. The smaller machines are hydrostatically driven, while the five larger models, ranging from 4.0 to 8.1 cubic yards capacity, are mechanically driven through a power-shift electronically modulated transmission and torque converter. Deere's flagship loader is the 844K fitted with a John Deere 380-net-horsepower PowerTech™ diesel that meets EPA Tier 3/EU Stage IIIA emission regulations. Operating weight with standard bucket is 70,089 pounds. *John Deere*

The John Deere hydrostatic 755C crawler loader carries a bucket rated at 3.1 heaped cubic yards, and is powered by a Liebherr engine rated at 177 net horsepower. The rear-mounted engine, as found on most modern crawler loaders, gives perfect balance to the loaded bucket, and better visibility for the operator than earlier front-mounted engine types. The single heavy-duty bucket cylinder give maximum pry-out force through the Z-type linkage.

Doosan from South Korea has come to prominence in recent years with the purchase of a number of well-established earthmoving equipment makers including Bobcat and Daewoo. The latter's loaders, now sold as Doosan machines, are represented here by the DL500 working in a concrete recycling yard fitted with a 6.8-yard bucket. The machine has a Cummins engine of 335 net horsepower and weighs 66,017 pounds.

As previously mentioned in the crawler tractor section, Dressta machines are built in Poland by Huta Stalowa Wola (HSW) and exported around the world. The 560E 'Extra' is the company's current largest wheel loader, and carries a general purpose bucket of 8.5 cubic yards. Power comes from a 427-horsepower Cummins diesel which meets EPA Tier 3 and EU Stage IIIa emission regulations. The 560E is a direct descendent of the former American-built International Hough and Dresser machines. *Marek Stankowski, HSW*

This mid-size Hitachi ZW310 is one of a dozen different sized wheel loaders made by the Japanese company and distributed around the world. It features Z-bar linkage for parallel lifting, advanced automatic transmission and computerized torque control system. The 51,700-pound loader works with bucket capacities ranging from 4.2 to 6 heaped cubic yards depending on the material, and its engine puts out 295 horsepower. *Andreas Barner*

Headquartered in England, JCB concentrates on the smaller end of the construction equipment market but sells huge numbers of machines from its factories located around the world. The 436 HT is equipped with a tool carrier featuring parallel bucket lifting and hydraulic quick-hitch attachment changing, which can be done from the operator's seat. The 436 HT employs a 165-net horsepower Cummins engine and tips the scales 30,630 pounds in standard configuration.

A Kawasaki 65Z IV loader is the maid-of-all-work on this sewer replacement job in San Diego, California. With its "quick-hitch" attachment, different bucket types and forklift arms can be switched in a matter of seconds without the operator leaving his cab. The 134-net horsepower engine supports a 2.6 cubic yard standard bucket.

On city construction jobs wheel loaders can carry out a multitude of tasks from placing gravel and carrying materials, to backfilling trenches and loading trucks. This Komatsu WA380-5 works on a road job in Edmonton, Alberta. Heaped bucket capacity is rated at 4 cubic yards, the Komatsu engine develops 187 horsepower, and the machine weighs 38,580 pounds.

Komatsu's WA1200-6 top-of-the-line wheel loader is targeted for production in surface mines. At the time of writing it's the world's third largest wheel loader, with two larger models available from LeTourneau. With 40-ton payload, the standard rock bucket holds 26.2 heaped cubic yards, and machine service weight is 476,000 pounds. Power comes from a 16-cylinder Komatsu diesel engine of 1,560 horsepower, driving through a power shift transmission with automatic monitoring of engine speed to suit digging conditions. *Komatsu*

LeTourneau's wheel loaders are designed for the surface mining industry. They are all diesel-electric drive with motors contained in each wheel hub. The range begins at a size way beyond the upper limits of most manufacturers. The smallest, the L-1000 is shown here working in a coal mine in Alberta, Canada, with a standard 17-yard bucket. The engine powering the advanced solid-state computerized control system is rated at 925 horsepower.

Poised with 28 cubic yards of earth and rock, waiting for the next truck to arrive, this LeTourneau L-1400 loader is an overburden production machine at an Ohio surface coal mine. The advanced computer-controlled electric drive system is powered by a 16-cylinder Cummins diesel rated at 1,800 horsepower. Because of the high cost of tires for these big loaders, chains are often used on the front set to protect from cuts and abrasion in rocky conditions.

The LeTourneau L-2350 is the largest wheel loader ever built. This picture shows it comfortably loading a 400-ton capacity Liebherr T282B, the largest-size truck available today. The gargantuan loader wields a standard bucket rated at 53 cubic yards or 80 tons payload, and if the machine could be run on weigh scales large enough, they would read 586,000 pounds. In 2011 Joy Global (P&H) announced it had purchased LeTourneau Inc. including its line of wheel loaders. *Eric C. Orlemann*

An operator inspects the 70/70-57 13-foot, 6-inch diameter tires after his shift on this LeTourneau L-2350, the world's largest wheel loader. The twin exhaust stacks conceal a 16-cylinder Detroit Diesel engine rated at 2,300 horsepower. While cost-per-ton loaded by a big wheel loader is higher than similar-sized electric shovels, the loader's superior mobility makes it an essential tool in any large surface mine.

As with its crawler tractors, Liebherr wheel loaders are hydrostatically driven. Here is the L 586, the company's flagship loader, working in a German quarry. It wields a standard bucket holding 7.2 cubic yards, and weighs 54,200 pounds in operation. The engine is Liebherr's own D936L A6 diesel putting out 340 horsepower. The engine output shaft faces to the rear, allowing the hydraulic pumps to act as a counterweight, resulting in higher tipping loads.

Liebherr's hydrostatic crawler loader line is represented here by the LR 641. With a standard bucket of 3.8 cubic yards and operating weight of 53,900 pounds with standard bucket, the loader sports a Mercedes-Benz diesel rated at 219 horsepower. Hydrostatic drive produces smooth, stepless control with independent drive to each crawler, enabling power and spin turns.

New Holland equipment today is marketed by New Holland Kobelco Construction Machinery S.p.A., one of the Fiat Group. Shown here in a tough rock application with chain tire protection is the W270B, one of the company's wheel loaders. It operates with a standard bucket of 5.4 cubic yards, and weighs in at 51,740 pounds. A Cummins 296-horsepower engine drives the machine through a torque converter and power-shift transmission. *Andreas Barner*

Over several decades, Sweden's Volvo has built a solid reputation in wheel loaders, specializing in this product for many years. This L350F from Volvo's current line is powered by a 528-horsepower Volvo electronically-controlled, low-emission engine which delivers full power even at low revolutions. Drive is through an automatic power-shift transmission with torque converter. Standard bucket carries 9 cubic yards, and machine service weight is approximately 110,000 pounds. *Urs Peyer*

CHAPTER 3: SCRAPERS

The 657G is the world's largest motor scraper available today. Built by Caterpillar, the scraper holds 44 heaped cubic yards, and when equipped with Caterpillar's patented "Push-Pull" system, weighs 157,750 pounds empty. The front and rear engines combine to produce a maximum net 1,051 horsepower. *Urs Peyer*

Self-propelled scrapers are spectacular digging and hauling machines that collect a load of earth, carry it to the fill, and return at speeds up to 30 miles per hour. These machines consist of a tractor unit coupled to a scraper bowl or bucket, an apron to close over the front to retain the load, and a powered ejector plate that pushes the load out of the front when the apron is raised. Some types sport an additional engine on the rear to provide four-wheel drive when needed in soft conditions or hauling up steep grades.

During the loading part of the cycle a scraper is usually assisted by a bulldozer pushing behind. To a limited extent, the double engine types are able to load themselves, but a push tractor is always recommended to reduce tire wear and increase productivity. There are, however, some true self-loading scrapers which are able to efficiently load themselves. These are elevating scrapers, auger scrapers, and "push-pull" types where two self-propelled scrapers are hooked together during the loading cycle. These types are explained later in this chapter.

Until the 1970s, cable-operated scrapers pulled behind crawler tractors were popular, but became outmoded and replaced by easily operated hydraulic scrapers. In the past decade or so, pull-type scrapers have staged a comeback due to the advent of high-horsepower four-wheel-drive farm-type tractors, and multi-track tractors. Able to haul two or three scraper units at a time, these scrapers offer another high-speed alternative for the earthmoving contractor.

Cat's Push-Pull system is demonstrated here on a pair of Caterpillar 637G scrapers. To apply the Push-Pull system, two twin-engined scrapers are hooked together during loading to allow the power of all four engines to be applied to the cutting edge, in this case 1,566 horsepower for the 637G's. The scrapers are loaded one at a time. When filled, they unhook and operate independently on the haul, dump and return parts of the cycle.

Here's an example of an elevating scraper, sometimes known as a "paddle wheel" scraper. It's a true self-loading machine as it does not need a push tractor to assist loading. It utilizes a set of hydraulically-powered elevator flights to chop the material and hoist it into the bowl. The Caterpillar 623G heaps 23 cubic yards, and runs with a 3406E diesel of 272 net horsepower.

Another type of self-loading scraper is the auger scraper which carries a vertical, hydraulically-driven auger in the bowl. The auger draws the material upwards, filling the bowl to its heaped capacity. This allows the full tractive power of the scraper wheels to be applied to cut the material and draw it into the bowl, thus reducing loading time and fuel consumption. This Caterpillar 657E Auger scraper shares the same heaped capacity and engine horsepower as the standard version.

A Caterpillar 651E empties its load on a topsoil salvage pile. Powered by a single 605-net horsepower Caterpillar 3412E engine, the 651E holds 44 heaped cubic yards in its bowl. All Caterpillar scrapers feature a Cushion Hitch, a set of hydraulically cushioned link rods in the tractor/scraper articulated joint that dampens the vertical bounce associated with high-speed travel, resulting in a smoother ride.

Some contractors still prefer tractor-drawn scrapers for certain applications. As hydraulics became more reliable, easier to operate and maintain, many old cable winch, pull-type scrapers were converted to hydraulic operation. This 300-horsepower Caterpillar D8K tractor makes a perfect digging tool when matched to a Caterpillar 435 scraper with 18 cubic yards heaped capacity.

With a pair of John Deere 1810C scrapers in tow, each carrying 18 heaped cubic yards, this John Deere wheel tractor maintains the pace in this earthmoving application. John Deere offers a three-machine line of "Scraper Special" wheeled tractors ranging from 425 to 530 net horsepower, tailor-made for pulling up to three scrapers. The 1810C has a cutting width of 10 feet, and an empty weight of 20,500 pounds.

Designed to be pulled by a 6-wheel drive, 40-ton articulated dump truck (ADT) in the 450-horsepower class, with box removed, the K-Tec 1254 ADT scraper carries 54 heaped cubic yards, making it the largest pull-type scraper available on the market. The scraper gooseneck attaches to a "fifth wheel" on the tractor unit which is positioned to provide optimal weight distribution on the outfit's four axles when loaded. This type of tractor unit also provides a smooth stable ride for speeds up to 30 miles per hour. *K-Tec*

Able to move dirt at almost the same speed as their wheeled counterparts, tractors equipped with four flexible rubber track assemblies instead of wheels are designed to work well in soft conditions. Bell offers its standard 413 net horsepower 4206D articulated wheeled tractor unit converted to rubber tracks. It is capable of pulling up to three 18-yard scrapers and employs a modular track system developed over a 10-year period by ATI in conjunction with Goodyear Tyre & Rubber Co.

CHAPTER 4: GRADERS

Caterpillar's H-series motor graders announced in 1995, and its updated M-Series launched in 2007, maintain the company's position as no. 1 grader manufacturer. On a tough assignment in stiff clay, a late-series 14H pulls through the job without hesitation. This 14H is equipped with 8-speed power shift transmission and torque converter providing variable power: 220 net horsepower in gears 1 to 3, and 240 net horsepower in gears 4 to 8. The fully equipped machine tips the scales at 58,509 pounds.

Motor graders are probably the most familiar earthmoving machines in North America. Seen on highways clearing snow in winter or maintaining gravel roads in summer, the grader is often the lifeblood of our transportation system. Today, the largest application for graders is still county and municipality road maintenance, but a close second is their use in major earthmoving projects including road construction and surface mining.

The grader carries a centrally mounted blade or moldboard which has many independently-controlled operating positions. The blade is raised or lowered independently at each side allowing a side slope or cross-fall to be shaped as the grader moves forward. The blade can rotate horizontally so that material can be pushed to either side. Also the blade's pitch can be changed to cut into the material, or simply push it along the road surface. The blade can move sideways to reach out far beyond the wheel track, and when trimming side slopes, can be positioned at any angle from horizontal to vertical.

This 160M from Caterpillar's current M-series demonstrates precise control in city working conditions. A Caterpillar C9 engine provides variable power from 213 to 248 net horsepower depending on the gear selected. Caterpillar's M-series graders feature two joystick controls with thumb buttons for all grader and blade movements, eliminating the bank of levers traditionally found on older graders. The 160M weighs up to 52,795 pounds with standard 14-foot blade.

The Caterpillar 16M is the standard for haul road maintenance on all but the very largest earthmoving projects and surface mines. With its 16-foot blade and variable-horsepower C13 engine (297 – 332 hp), the 16M keeps pace with large haul truck fleets at speeds up to 33 miles per hour. Fully equipped, the 16M weighs 54,570 pounds.

The world's largest haul trucks need the world's largest graders. When working with haul trucks 320 tons capacity and larger, the grader becomes a production tool, ensuring smooth roads to protect truck tires and suspensions. First with its 16-series and, since 1996, with its 24-series graders, Caterpillar has dominated the large grader market, with virtually no competition. The 145,808-pound 24M sports a 24-foot blade which it pushes with a 533-net horsepower C18 engine.

This 780A was the flagship grader of the former line built by Champion Road Machinery Ltd., purchased by Sweden's Volvo in 1997. The 46,100-pound machine was available in tandem-drive, or all-wheel 6x6 drive with hydrostatic drive to the front wheels. The 780A featured a variable horsepower arrangement with its Cummins engine putting out 210 flywheel horsepower in 1st and 2nd gears, and 235 flywheel horsepower in gears 3 to 8. Champion's compact grader line was sold to a successor Champion company in the USA who still markets them today.

John Deere's line of modern motor graders is topped by the 48,620-pound model 872G/GP. The "GP" stands for GradePro, which means the grader will accept an integrated electronic grade-control system of the owner's choice. The Deere electronically-controlled engine and eight-speed PowerShift Plus™ transmission provide variable power from 221 to 283 horsepower. Hydrostatically driven front wheels provide six-wheel drive through a power-management system to balance traction between the front and rear wheels. *John Deere*

Here is Komatasu's GD670A-2 grader with 204 net horsepower and 32,400-pound operating weight. Today, most graders boast articulated frames and, in addition to all the usual blade movements, are equipped with powered leaning front wheels to offset the side load produced by the angled blade. Recently, Komatsu upgraded this model to the GD675-3, with variable power from 180 to 200 net horsepower, and weight upped to 34,855 pounds.

Not all graders have to be big to be useful. A number of manufacturers produce small 'compact' graders that perform an ever-increasing list of tasks on any construction job. Representing this class of machine is a NorAm 65E Turbo grader, helping to re-pave the world-famous Las Vegas 'Strip.' The machine carries a 10-foot blade and weighs 16,800 pounds. A 110-horsepower Cummins QSB4.5 engine provides the power.

Volvo enhanced its heavy equipment lines when it purchased Champion, one of the world's oldest grader manufacturers, in 1997. The current G-series includes the G976, claimed by the makers to be the largest 6-wheel drive grader available today. The Volvo D9B engine and variable power system provide 225 to 265 net horsepower depending on one of three ranges selected. With standard 12-foot blade the G976 weighs 43,650 pounds. *Volvo*

CHAPTER 5: HYDRAULIC EXCAVATORS

One of the most extensive ranges of hydraulic excavators is offered by Caterpillar. Machines range from the 3,792-pound model 301.6 to the 190,840-pound 385C L, with over 30 different sizes offered, including five wheel-mounted models. The 385C LME is the "Mass Excavation" version, seen here at a British coal mine uncovering coal at the surface. The LME version uses a 23-foot, 9-inch boom and 7.5-cubic-yard bucket, compared with a longer 27-foot, 7-inch boom and 6-yard bucket on the standard excavator.

Hydraulic excavators work on every kind of construction job from road maintenance, trenching and foundation work to mass excavation on major industrial sites, as well as in quarries and surface mining operations. Machines range from the 13 horsepower 300.9D to the 523 horsepower 390D, with over 30 different sizes offered including five wheel-mounted models. They are made in every industrialized country by hundreds of manufacturers around the world. Although special application attachments are available, the vast majority are deployed as front shovels or backhoes. A typical hydraulic excavator is powered by a diesel engine driving a series of hydraulic pumps which, at the control of the operator, transmit oil under high pressure via flexible hoses to hydraulic cylinders and motors for the excavator's motions.

Hydraulic excavators have largely replaced cable-operated types except in the very largest sizes. Certainly in small and medium sizes they are easier to operate, travel faster, and have positive action in all movements instead of relying on gravity to provide some of the digging forces. The largest sizes have gradually penetrated into major surface mines of the world where they work alongside similar-sized electric rope-operated shovels to load the world's largest off-road haul trucks.

Caterpillar's 5000-series mining excavator line was topped by the 350-ton model 5230. It dug with a 22-yard bucket, and a 1,470-net horsepower Cat 3516 diesel supplied the power. In 2003, Caterpillar withdrew from the hydraulic mining excavator market in anticipation of bigger and better things to come. In November 2010, the company announced its takeover of Bucyrus International, giving Caterpillar access to the former O&K and Terex machines of up to 1,000 tons operating weight.

Most modern hydraulic excavators can be specified with different boom and stick lengths to suit special applications. This Caterpillar E300B extreme-reach backhoe is ideal for cleaning out ponds such as this one in Arizona. The machine, in the 70,000-pound weight class, is powered by a Caterpillar 207-net horsepower engine. Cat's E-series excavators resulted from a joint venture with Mitsubishi of Japan.

In 2000, Case introduced its state-of-the-art CX line, represented here by the CX-700B equipped with a 463-net horsepower Isuzu engine. The CX700B features three work modes: Automatic, Heavy and Speed Priority. Auto Power Boost also increases standard bucket digging forces by 10 percent. Maximum digging depth is 32 feet, and working weight is 151,896 pounds. *Case Construction Equipment*

This Deere 800C excavator demonstrates its stability on bank slope trimming. The 168,540-pound machine is powered by an Isuzu engine rated at 396 net horsepower, and can handle a standard bucket holding 4.6 cubic yards. Since 1988, a joint venture with Japan's Hitachi has resulted in all Deere excavators today built to Hitachi designs.

As a world leader in hydraulic excavators, Hitachi is one of the few manufacturers venturing into the "super-class" mining excavator sizes up to 800 tons. The EX1900, shown here at an Alberta oil sands operation, waits for the next truck with a heaped load of 15 cubic yards. The EX1900-6 sports a Cummins QSK38-C engine producing 1,039 net horsepower to supply a hydraulic system at 4,270-pounds/square inch pressure. Backhoe operating weight is 423,280 pounds.

An Indiana coal mine uses this Hitachi EX3600 excavator to load its fleet of Euclid trucks. In Hitachi's mining excavator range, the EX3600, with an operating weight of 795,900 pounds, is only a mid-size machine. Bucket range is 27 to 30 cubic yards, and the big Cummins QSK60-C engine supplies 1,944 net horsepower.

Hitachi's EX8000 is its biggest excavator to date. Able to comfortably load 240-ton to 320-ton haul trucks, its bucket holds 52 cubic yards. The house measures 35 feet wide and height to top of the cab is 32 feet, 6 inches above ground. It takes two Cummins engines to produce the necessary 3,880 horsepower to drive the hydraulic system. If scales could weigh this monster, they would read 1,787,900 pounds.

Showing superior stability, a JCB JS290LC thrills the crowd at a U.K. equipment demonstration. With a 216-horsepower Isuzu engine, the 68,000-pound machine can dig down to 23 feet with 1.6-cubic-yard bucket and standard boom and arm. JCB crawler excavators are equipped with JCB's Advanced Management System (AMS) which matches engine and hydraulic outputs to working demands. AMS also records key data that can be downloaded to a computer so problems can be diagnosed.

Since the 1970s, mini-excavators have become extremely popular for every small excavating job, taking over work previously done by hand labor. All the major excavator manufacturers offer these small machines, including many that specialize in mini-excavators (machines less than 10 tons in weight.) The JCB 8080 at the top end of the company's popular line of mini-excavators weighs 18,500 pounds and its engine develops 58 horsepower.

Komatsu makes a full line of excavators from the smallest to some of the largest in the industry. Making full use of its superior reach, a Komatsu PC600LC helps prepare foundations for a light rail transit system in Edmonton, Alberta. The current PC600LC-8, in the 130,000-pound weight class, carries a standard bucket of 3.4 cubic yards, and its Komatsu engine puts out 429 net horsepower.

Along with Liebherr, Hitachi, and Bucyrus/Terex, Komatsu competes in the big league of hydraulic mining excavators since taking over the former Demag line from Germany. This PC5500 loads a 240-ton Caterpillar truck in the Alberta oil sands mining operations. Two Komatsu engines with a combined 2,520 net horsepower provide the necessary power to push the 37-yard bucket through the bank. Machine weight is listed at 1,210,000 pounds with 71-inch track shoes.

The massive PC8000 is the biggest excavator in Komatsu's roster. This one keeps busy loading 320-ton Komatsu 930E trucks with its 55-yard bucket at an Alberta oil sands operation. As usual on today's large machines, a computer monitoring and diagnostic recording system provides data on machine performance and diagnostic fault detection. The PC8000 operating weight is listed at 1,565,000 pounds, and two Komatsu diesels totaling 4,020 horsepower provide the power.

Short or zero tail swing radius excavators have recently increased in popularity. They are designed for working in close quarters next to structures, or in narrow traffic lanes without impacting adjacent traffic. Komatsu offers its 25-ton PC228USLC-3 in this category which sports a Komatsu 143-flywheel horsepower engine. Stability is maintained by the machine's long and wide undercarriage.

This Liebherr 974 looks none the worse for wear after many years of tough work in a hard rock quarry in Belgium. The 150,000-pound machine is shown placing another load onto a 50-ton Payhauler four-wheel-drive hauler. The 974 backhoe has a digging depth of just under 28 feet when equipped with a 13-foot stick. Under the hood is a Cummins 450-horsepower engine.

This Liebherr 995 works in an Ohio surface coal mine loading blasted overburden into Liebherr trucks. It wields a 34-yard bucket, and the hydraulics exert a maximum crowd force of 202 tons. The engine is an MTU diesel developing 2,140 horsepower, and machine weight is 992,080 pounds. The 995 is also available as a 34-yard backhoe.

In 2010, Liebherr commissioned its flagship R9800 hydraulic mining excavator at Burton Coal, Queensland, Australia. Billed as the largest backhoe excavator in the world, this machine is operating with a 58-yard bucket and loading 240-ton trucks in three passes. The R9800 is powered by two Cummins QSK60 diesels providing 4,000 horsepower, and operating weight is 1,763,630 pounds. The R9800 is also available as shovel with a 55-yard heavy-duty bucket. *Liebherr*

Current New Holland excavators are derived from the Japanese Kobelco line following a joint venture agreement set up between CNH Global and Kobelco in 2002. The E485B in the 52-ton class is available in deep digging or mass excavation options, with buckets from 2 to 3.4 cubic yards capacity. A 346-horsepower Hino engine provides the power. The New Holland features a "Continuous Power Boost" which increases hydraulic pressure at the command of the operator when tough digging is encountered. *Andreas Barner*

The larger machines of O&K's former extensive excavator line became part of Terex Mining Division in 1997. In 2009, Bucyrus took over Terex Mining, including Unit Rig trucks, and in 2011 Bucyrus became part Caterpillar. Smallest of the "mining" excavators is the RH 30-E with a service weight of 176,000 pounds, and engine output of 436 net horsepower. The machine shown here loads a 100-ton haul truck with a 6.5-cubic-yard bucket during the restoration phase of a British opencast coal site.

This stalwart O&K RH-200 excavator proves its longevity at an Alberta oil sands operation. Working round the clock, the 16-year old machine loads overburden into 320-ton haul trucks. Two Cummins KTA 38C engines power this RH-200 which carries a 32-yard bucket. Although more expensive to operate than electric powered rope-operated shovels, hydraulic diesel-powered excavators are essential additions to shovel fleets because of their superior mobility, and lack of trailing power cable.

The Terex RH-120E is the latest of a long line of 120-series excavators, going back to the first O&K RH-120 in 1983. This one works in a German cement quarry and carries a 22-yard shovel bucket. Working weight is 620,820 pounds and engine output 1,350 net horsepower. All O&K mining excavators exhibit the patented TriPower linkage which connects the bucket, boom and stick movements through a rotating trunnion mounted on the boom. Advantages include automatic constant bucket angle when crowding, and increased crowd and lift forces.

Here is king of all hydraulic excavators, the Terex/O&K RH-400, the world's largest and most powerful. Able to comfortably load the world's massive 400-ton haul trucks, the machine shown is one of seven shipped so far to the Alberta oil sands. The RH-400 works with a 57-yard bucket and tips the scales at 2,216,500 pounds. Two big Cummins QSK60-C engines produce 4,400 net horsepower. Operator's eye level is approximately 30 feet above ground, and machine undercarriage measures 28 feet, 3 inches wide. Under Caterpillar this model is referred to as the 6090 FS.

CHAPTER 6: ROPE-OPERATED CRAWLER EXCAVATORS

Three-pass loading a 320-ton Komatsu 930E haul truck is easy for this Bucyrus 495HF electric shovel with its 60-yard bucket. One of the largest shovels available today, the 1,482-ton 495HF has a 67-foot boom and 47-foot dipper handle carrying dippers from 40 to 80 cubic yards capacity depending on material density. Electrical equipment utilizes AC variable frequency control. The model designation for this machine under the Caterpillar regime is the 7495 HF.

In the earthmoving world, "excavator" refers to several different classes of digging machines. Most common today is the hydraulic excavator (Chapter 5), available in front shovel or backhoe form. But "excavator" can also refer to walking draglines (Chapter 8), continuous excavators (Chapter 9), as well as the crawler-mounted cable or rope-operated shovels and draglines in this chapter. One of the very earliest forms of mechanical excavators, they evolved from the part-swing railroad shovel of the 1830s, to the more sophisticated fully-revolving crawler shovels of the 1920s, to the world's largest mobile land machines of the 1960s.

Rope-operated crawler excavators were the most popular loading machines from the 1920s through the 1960s. But their popularity declined after that date as technology brought reliable hydraulic excavators to the fore. By 1980, the smaller and medium diesel-powered sizes were all but eclipsed by other types of equipment, such as hydraulic excavators and wheel loaders. But at the larger end of the scale, the boom in surface mining in the 1970s called for bigger rope shovels to load larger trucks, and this trend continues today with the largest shovels loading 400-ton capacity trucks in four passes.

Today, only two manufacturers serve the western world with large electrically powered shovels: P&H Global and Bucyrus International which was purchased by Caterpillar in 2011. Chief competitor of these large machines, the former Marion Power Shovel Co. of Marion, Ohio, was purchased by Bucyrus in 1997.

A supervisor poses against the 125-inch wide crawler tracks of a Bucyrus 495HF electric shovel in the Alberta oil sands. One can only compare the size of a man with a large mining machine when it is shut down. When working, personnel and passenger vehicles stay well clear! The HF stands for "High Flotation," and the super wide crawler shoes provide this capability when underfoot conditions are soft.

Another in the popular Bucyrus 495-series is the 495HR electric shovel. With HR indicating "Hard Rock," this shovel comes equipped with standard 79-inch shoes enabling it to work on a hard quarry floor. For over five decades, Bucyrus shovels featured a rope-operated crowd system. This places the crowd machinery on the deck to reduce swing inertia, and allows a lighter boom. The most recent 495-series shovels are available with a hydraulic crowd cylinder to replace the rope system. Under Caterpillar this model is referred to as the 7495.

Manitowoc's biggest dragline is the 6400 which carries a 15-yard bucket on a 160-foot boom. Optional booms up to 200 feet with a 10-yard bucket are also available. Height to top of cab is 23 feet, 9 inches, and machine working weight is approximately 600 tons. The 6400 employs a dragline interlock system where the hoist and drag drums are connected by a chain and clutch. When locked together during the hoisting phase, the pull on the drag rope as it is paid out reduces the hoisting effort required.

The Manitowoc 6400 boasts power from two big Cummins diesels. A 16-cylinder KTA-3067 at 1,600 horsepower drives the upper works, while a KT-150 at 450 horsepower in the lower works drives hydraulic pumps to provide independent drive to the crawler assemblies. Each 83-inch wide crawler is gear driven by two hydraulic motors through planetary and spur gear reductions, permitting smooth turns and the ability to counter-rotate the tracks.

The stalwart P&H 2800 series electric shovels, introduced in 1968, have been the mainstay of many surface mines around the world. This 2800XPB is shown loading 200-ton LeTourneau Titan trucks at an Arizona copper mine. Nominal dipper capacity is 46 cubic yards, boom length is 58 feet, and machine operating weight is approximately 1,130 tons. A P&H patented DC "Electrotorque" system powers the machine.

This P&H 4100XPB electric shovel is a key player in a Wyoming coal mine, loading a variety of trucks up to 400-ton capacity with its 80-yard dipper. The 4100 series electric shovels from P&H raised the bar in size and power when unveiled in 1991. Since then, a dipper capacity increased to 100 tons from the original 85 tons, and further advanced electrical systems resulted in the first 4100XPB in 1999.

The P&H 4100 BOSS (Series B Oil Sands Special) was specially designed to work in the Alberta oil sands. Although slightly less in working range, weight and power than its XPB cousin, the BOSS boasts extra wide 138-inch crawler shoes, reducing the ground-bearing pressure of the 1,500-ton shovel by 45 percent to 30.8 psi., low enough to operate on the soft oil-drenched sands.

Another P&H 4100 BOSS shovel heaps a load on this 240-ton Caterpillar 793B hauler in two or three passes. Both the 4100XPB and 4100 BOSS employ the patented P&H Electrotorque control with static DC power. With his eye level at 32 feet above ground, the operator has a clear view of the action. Touch screen monitors in his cab keep track of all machine functions as well as provide diagnostic information to rapidly identify faults.

In 2008, P&H announced its C-series electric shovels incorporating latest electrical technology in its entire line of shovels, draglines and blast hole drills. Although exhibiting similar ranges and capacities to their predecessor models, C-series machines operate with P&H Digital Drive control and are offered with either DC or AC drive technology. The special oil sands shovel version became the 4100C BOSS with nominal dipper capacity of 63 cubic yards. The one shown is equipped with DC electric drive.

An AC-drive P&H 4100C BOSS loads a Caterpillar 793D hauler with 250 tons of overburden. The undercarriage on this oil sands special shovel with 138-inch shoes measures 41 feet, 11 inches wide and 38 feet long. Overall house length including counterweight is 56 feet, 5 inches. This shovel, and its companion 4100XPC hard rock version with an approximate weight of 1,690 tons, are the world's largest shovels working today.

P&H developed some large two-crawler draglines based on its electric shovels. In 1981, the company introduced its largest crawler dragline, the model 2355. This state-of-the-art, two-crawler machine was supplied with either diesel or electric power. The diesel version carried a pair of diesel engines totaling approximately 2,000 horsepower. The electric version shown has an operating weight of 765 tons, and swings an 18-yard bucket on a 160-foot boom. It removes parting material in a Saskatchewan coal mine.

An alternative to mechanical or electric drive is hydraulic drive. Some of the larger cranes produced by Germany's Sennebogen can be used as draglines. This 6180HD, working in a clay pit in central England, has a 100-foot boom carrying a 6-yard bucket. In this application, a non-standard AC electric motor rated at 600-horsepower replaces the normal diesel engine to drive hydraulic pumps for all the machine's motions. Hoist, drag, and swing motions, as well as boom hoist and crawler propel, are driven by variable displacement hydraulic piston motors through planetary gears.

CHAPTER 7: OFF-HIGHWAY TRUCKS

In 1934, the Euclid Road Machinery Company became the first to specialize in off-road haulers. Succeeding decades brought a rich heritage of heavy-duty haulers including some of the world's largest. After passing through several different owners, the company today is owned by Japan's Hitachi which builds the current EH-series Euclid-Hitachi haulers. This diesel-electric EH4000 with General Electric wheel motors is rated at 250 tons capacity. It is powered by a 16-cylinder Detroit Diesel developing 2,458 flywheel horsepower. *Eric C. Orlemann*

Off-highway trucks evolved from the need to haul vast amounts of material on unpaved roads. From their early beginnings, when attempts were made to adapt highway trucks to off-road use, these vehicles have gradually increased in strength and capacity. Now the largest behemoths carry loads of over 400 tons with a gross vehicle weight of over 1,300,000 pounds. These are the true monster trucks of today.

There are several different types of off-highway trucks or haulers built for different applications. These include articulated dump trucks up to 50 tons capacity, rigid frame rear dumping and bottom dumping types, and tractor-trailer units with high capacity bodies for hauling coal.

Recent improvements in haulage vehicles include electronic controls and diagnostics, higher horsepower engines, more robust components and frame design, and smoother gear selection. Tilt and telescoping steering wheels are now standard equipment, as are well-situated controls and easy-to-read gauges. Today's operators expect air conditioning and sound suppression to reduce noise.

Large rigid frame haulers are built with either mechanical or diesel-electric drive. Recent technology improvements with electric AC drive have allowed manufacturers and their customers to favor this type over the former popular DC drive. The electric AC option is also making inroads into the mechanical drive market in larger size trucks, as customers discover efficiencies not possible in mechanically driven trucks.

Komatsu's large off-highway haulers trace their heritage back to 1953 when LeTourneau-Westinghouse Company was established. After several corporate changes, the line eventually became 100 percent owned by Komatsu. When the 930E hauler with 320-ton payload was introduced in 1996, it was the industry's first electric drive truck to employ AC electric wheel drive. Two versions are offered with Komatsu engines: the 930E-4 powered by a 16-cylinder diesel of 2,550 flywheel horsepower, or the 930E-4SE with an 18-cylinder diesel of 3,429 flywheel horsepower.

The latest version of Caterpillar's popular 777-series haul trucks is the 777F rated at 100 tons capacity. This one is discharging a load of overburden at a surface coal mine in England. The 777F carries a Caterpillar C32 engine putting out 938 net horsepower. The 7-speed electronically controlled planetary power shift transmission delivers constant power over a range of speeds up to 40 miles per hour.

Caterpillar's true monster truck shares title with the Liebherr T282C as the world's largest truck at 400 tons rated capacity. The Caterpillar 797B and the current 797F are mechanically driven through an electronically controlled planetary power shift transmission. Diesel engine for the 797F is a 20-cylinder Caterpillar C175-20 delivering 3,795 net horsepower to the power train. Overall truck width is 32 feet and height to canopy top is 25 feet.

Liebherr, founded in Kirchdorf, Germany, entered the large mining truck business in 1996 when it took over the American Wiseda line of diesel-electric drive trucks. Working as a team in an Ohio surface coal mine, this Liebherr T252 truck is loaded by the 34-yard clamshell-type bucket of a Liebherr R995 hydraulic excavator. The 200-ton capacity T252 truck has an empty weight of 327,000 pounds, and is powered by either a Detroit Diesel or Cummins engine in the 2,000-horsepower class.

The Liebherr T282B at 400 tons capacity jointly shares title with the Caterpillar 797F as the world's largest truck. The electric AC drive T282B and the recently-introduced T282C offer a choice of engines in the 3,500-horsepower class: a 20-cylinder MTU, or an 18-cylinder Cummins. The T282C, introduced in 2010, sports an improved cast steel frame and a redesigned rear axle box with two service doors. The AC electric drive system is designed and built by Liebherr.

Which end is the front? This is a unitized rigid-frame bottom-dump coal hauler, a concept introduced by Kress Corporation in 1970. When loaded, the vehicle weight is carried equally on eight wheels, mounted in pairs at each corner. The straddle-mounted front wheels can turn 90 degrees to the frame for tight turns. A rear-mounted engine drives the rear wheels through a mechanical transmission. The 160-ton CH-160 illustrated hauled coal at a Saskatchewan coal mine, and utilizes a 1,200-horsepower Detroit Diesel engine.

This Ohio coal mine uses a LeTourneau L-1400 wheel loader to load a fleet of 260-ton Terex Unit Rig MT4400 trucks. Unit Rig brought the electric wheel drive concept to the forefront in the 1960s, capturing the lion's share of the big hauler market. As part of the Terex organization since 1988, taken over by Bucyrus International in 2009, and by Caterpillar in 2011, the Lectra-Haul line continues to compete in the largest haul truck size classes.

A Terex Unit Rig MT4000 truck dumps 240 tons of coal into the receiving hopper at a Wyoming coal mine. The truck is equipped with a Phillipi-Hagenbuch "HiVol" high volume body to maximize its carrying capacity. It also boasts an "Autogate," a tailgate patented by Phillipi-Hagenbuch that moves clear of the load when the truck body is raised.

Moving overburden at a Wyoming coal mine, this Terex Unit Rig MT5500 dumps 360 tons in the worked-out pit. The MT5500 features a unique AC electric drive system that allows the 600-ton fully-loaded truck to move from full propel to full retard in less than a second, and come to a complete stop by use of dynamic retarding. The latest MT5500B version can be ordered with a choice of Cummins or MTU/Detroit diesel engines from 2,700- to 3,500 flywheel horsepower.

For long hauls on relatively flat ground, the tractor-trailer configuration for coal hauling is most economical. A rigid-frame haul truck, when converted to hauling a trailer with a fifth wheel coupling, can usually haul between 50 to 70 percent more tonnage than its normal rated carrying capacity. This Euclid CH-160 hauler, based on a Euclid R-85B tractor unit, would normally carry 92 tons as a rear dump truck, but with this trailer built by Mega Corporation, it hauls 160 tons.

Haulmax presents earthmovers with an alternative haulage option when faced with long hauls. According to the makers, the vehicle's multiple axle design, superior braking characteristics, and lower initial cost, make one-way hauls up to 30 miles long economical. Developed jointly with Caterpillar, and designed and built in Tasmania, Australia, Haulmax vehicles use an integrated Caterpillar power train, including engine, transmission, final drive and wet disc brakes. The Haulmax 3770D shown is powered by a 730-net horsepower 3412E engine for a payload of 80 tons. *Gavin Handley*

The articulated dump truck (ADT) originated in England in the late 1950s, but it was Volvo of Sweden who promoted and made it popular in the 1960s. The ADT is a unitized tractor-trailer vehicle with an articulated frame and rear-dumping body. This Volvo A35C receives a load from a Kobelco SK300LC IV backhoe on a golf course construction job near Calgary, Alberta. The Volvo ADT engine outputs 322 net horsepower and carries 35 tons.

This Volvo A40D articulated dump truck (ADT) carries 40-ton loads at speeds up to 34 miles per hour. Its drive train includes a torque converter and Volvo-designed 6-speed electronically controlled automatic planetary transmission, driven by a Volvo diesel producing 414 net horsepower. On Volvo ADTs, the driver is able to select four-wheel or six-wheel drive, providing 6x4 or 6x6 drive where conditions demand.

Two of Caterpillar's articulated dump trucks, the 30-ton 730 and 40-ton 740 are available with an ejector body option where the load is pushed out over the rear. Although sacrificing a slight reduction in carrying capacity, advantages include clean ejection of sticky material, on-the-go dumping, and superior stability at uneven dump sites. The top-of-the-line Caterpillar 740 Ejector model has a 453-net horsepower C15 diesel under the hood.

Liebherr announced its first articulated dump truck at the German Bauma show in 2010. With a 6x6 driveline, automatic transmission and torque converter, the TA 230 has a carrying capacity of 33 tons. Power comes from Liebherr's own D96LA6 diesel engine developing 367 horsepower. According to the manufacturer, a key design parameter for the TA 230 was compactness, as it exhibits transport dimensions similar to a typical 27-ton truck. *Urs Peyer*

Terex Unit Rig developed this 400-ton capacity MT6300 to challenge the 'world's largest truck' title currently claimed by Caterpillar and Liebherr. Shown on test in the Alberta oilsands, the AC electric drive truck features triple-reduction wheel motors with high gear ratios, an advantage when pulling away from the shovel in soft conditions. A 3,750-horsepower MTU/Detroit diesel provides the power. In 2009, Bucyrus took over Terex Mining Division, including Unit Rig trucks, and in 2011 Bucyrus became part of the Caterpillar organization.

CHAPTER 8 – DRAGLINES

This Marion 8750 walking dragline is kept busy uncovering coal at an Alberta, Canada, coal mine. Its 400-foot boom at 27-degree angle gives its 100-yard bucket an operating radius of 384 feet. It boasts eight hoist motors, six drag motors, six swing motors and four propel motors, for a total of 25,000 horsepower. Marion's "Tri-Structure" at the front of the house serves as gantry, mast and fairlead support. Operating weight is 6,915 tons.

Walking draglines belong to a family of excavators which includes some of the most massive machines ever to move on land. These monster machines work in surface mines where they move vast quantities of earth to uncover coal and other valuable minerals. Where the excavated material can be deposited within the range of the dragline, and where geology permits, the most efficient machine for bulk excavation is usually a dragline.

A walking dragline looks like a large crane, but instead of carrying a hook to lift loads, it carries a digging bucket suspended on the end of its boom by hoist ropes. In action, the bucket is dragged towards the machine to collect its load by another set of ropes, called drag ropes. When full, the bucket is hoisted, the machine swings, and the load is dumped in a pile to the side. To swing, the machine revolves on a circular base or tub on which the machine sits while digging. Except for some small models now obsolete, all walking draglines are electrically powered.

An important advantage of a walking dragline is the very low ground pressure it exerts because of its large-diameter base. The dragline often works on top of the material it digs, and soft material could not support the higher ground pressure of other excavators. A dragline's walking system is very simple. Two large shoes, one each side of the revolving frame, rotate simultaneously in a circular motion so that the shoes press into the ground, tilting the machine and moving it backwards one step for each shoe rotation.

Today, only two walking dragline manufacturers supply the entire western world with machines: Joy Global (P&H), and Bucyrus International which was purchased by Caterpillar in 2011. In 1988, P&H purchased Page Engineering Company of Chicago, originators of the dragline, to add this product to its roster. The only two other former walking manufacturers, Ransomes & Rapier Ltd. of England, and Marion Power Shovel Company of Marion, Ohio, were purchased by Bucyrus in 1988 and 1997 respectively, reducing world suppliers from four to two.

The front view of another Marion 8750 showing the "under-and-over" fairlead arrangement and operator's cab. This machine, last to be built by Marion Power Shovel before the company was taken over by Bucyrus International in 1997, carries 106-yard bucket on a 420-foot boom. The first 8750 dragline went to work in 1972, and that same nomenclature is used today by Caterpillar/Bucyrus for its current largest draglines. But, except for outward appearance, their design, capacity and range bear no resemblance to the earlier machines.

Another 125 cubic yards are dumped by this Bucyrus-Erie 2570-W at an Indiana coal mine. With a 335-foot boom and operating weight of 6,238 tons, the dragline's main DC motors total 26,100 horsepower. The 2570-W employs Bucyrus' patented "cam-and-slide" walking arrangement where an eccentrically-mounted wheel rotates inside a large roller bearing. The bearing and its frame slide on a greased path on the shoe when the wheel is rotated, producing the necessary circular motion to propel the machine.

Welding needed on this 125-yard bucket brings a maintenance crew and truck equipped with welder, tools and hydraulic boom crane. Even the smallest pin or chain link is a fair load to lift and maneuver into position. With a force of many hundreds of tons acting on the bucket teeth during digging, they get hot, wear, and have to be changed out and rebuilt regularly. But despite their huge size, bucket teeth are designed to be changed easily and quickly.

The Bucyrus line included the 680-W, a modular dragline with the advantage of fast erection time, and suitable for mines of a comparatively short duration. Components are simply bolted together on site in a matter of weeks, instead of many months for regular draglines. The 680-W features a state-of-the-art controlled-frequency AC electric drive system with independent AC motors for all main motions. The 680-W pictured carries a 25-yard bucket on a 225-foot boom and weighs approximately 1,100 tons.

This dragline boasts the world's largest dragline bucket operating today. Named "Ursa Major," the Bucyrus-Erie 2570-WS works in a Wyoming coal mine, and swings a 160-yard bucket on a 360-foot long boom. The "Super" digging machine possesses eight hoist, eight drag, fourteen swing, and four walk motors totaling 35,820 horsepower. If scales could weigh this monster, they would read about 7,000 tons.

From the operator's cab on the Bucyrus-Erie 2570-WS "Ursa Major" we see the dragline working bench. The dragline has swung about 180 degrees and is about to deposit 160 cubic yards of earth excavated from the working face behind. The dumped material will be used by the bulldozer to level the bench so the dragline can move ahead. It will dig down about 150 feet below bench level to uncover the coal, throwing the material onto the spoil piles seen at right.

With a 77-yard bucket, this Bucyrus 8200 uncovers coal in North Dakota. It was one of the first Marion-designed draglines shipped by Bucyrus after it took over the former Marion Power Shovel Co. This 8200 was specified with a 355-foot boom and six each of hoist, drag and swing motors, plus four walk motors for a total of 25,070 horsepower in 1997. Machine weight is quoted as 5,210 tons.

This Bucyrus 8750B dragline claims two world records: It is the world's heaviest dragline working today (8,068 tons), and boasts the longest boom ever installed on a dragline (435 feet). The boom angle at 32 degrees enables the 110-yard bucket to operate at 400 feet radius. The main motors have advanced AC drive and total 37,300 horsepower. Notice the two people outside the entrance door behind the cab. The machine is located at an Alberta coal mine.

When Joy Global (P&H) took over the former Page series of walking draglines in 1988, the line was completely modernized, even down to a redesigned walking system employing a cam-and-roller bearing. P&H also introduced its 9000-series draglines including the model 9020, largest yet built by the company. The 9020 in the picture is working at a surface coal mine near Estevan, Saskatchewan. It carries a 98-yard bucket on a 350-foot boom. Operating weight is 6,260 tons.

CHAPTER 9: CONTINUOUS EXCAVATORS

In 1980, Wirtgen of Germany began producing a line of continuous excavators known as Surface Miners. These machines were based on the company's already well-established line of pavement profiling machines. The crawler-mounted machines carry a centrally mounted drum with cutting teeth, and a rear-mounted conveyor which can swing through 180 degrees to load trucks moving alongside. The machine shown is the 800-horsepower 2200SM with operating weight of 112,000 pounds. *Wirtgen*

Continuous excavators move earth and other material in one continuous stream, unlike other types of earthmoving machines which operate in cycles of digging, moving and dumping. In the right conditions, continuous excavators attain a high level of efficiency because no interruption occurs in the excavation process. The largest continuous excavators are the giant bucket wheel excavators (BWEs) found in European open pit coal mines. These are the heaviest and most productive earthmoving machines of any type moving on land today.

BWEs represent some of the very largest excavating machines. They are found in surface mines and occasionally on large construction projects where vast amounts of material must be moved. BWEs excavate in a continuous manner utilizing a rotating wheel mounted at the end of a boom. The wheel is fitted with digging buckets around its circumference, and the buckets are filled by thrusting the rotating wheel into the working face while the boom swings from side to side. The excavated material falls onto a conveyor mounted within the boom structure which transports the material in a continuous stream on further conveyors to the discharge point. BWEs are usually connected to an elaborate system of conveyors which can transport the material a considerable distance. BWEs can also be arranged to load trucks, but keeping enough trucks on the haul road to take advantage of the BWE's full potential is the most challenging aspect of this method.

Geological conditions have to be just right to employ a BWE. Boulders are the machine's worst enemy, and when a breakage occurs, the entire earthmoving system grinds to a halt until the problem is fixed. This can be the major disadvantage of BWE systems when compared with discontinuous excavation systems such as shovel and truck spreads. But in the right conditions, nothing can match the tremendous output of a bucket wheel excavator used to its full capacity.

The Eagle 8300 is typical of a modern high-production wheel trencher. This trapezoidal type can dig a ditch eight feet deep and six feet wide at the bottom. Overall weight is 95,000 pounds and a 402-horsepower Caterpillar diesel provides the power. A laser-guided automatic steering and grade control system is standard. Since 1998, Eagle trenchers have been manufactured and marketed by Guntert & Zimmerman of Ripon, California. *Eagle Trencher*

This Trencor 1860HD is the "King of the Ditches" and claimed by its makers as the world's largest production-model trenching machine. The 225-ton brute digs trenches up to 35 feet deep and needs two engines to achieve its capabilities. A 1,200-horsepower Caterpillar 3512 engine has the exclusive job of driving the digging chain through a torque converter, while a 300-horsepower Caterpillar 3306B engine runs the crawler tracks and conveyors. *Tim Twichell collection*

Vermeer offers a full line of trenchers from the smallest to the largest, all hydrostatically driven. The 600-horsepower T1255 weighs 245,000 pounds and can operate as a regular ladder trencher or a terrain leveler as shown. Both types, fitted with rotary carbide-tipped teeth, are capable of milling through the hardest rock. The "terrain leveler" surface excavation machine can cut a path 12 feet wide, and is particularly suited to mining hard minerals such as limestone or gypsum.

The bucket wheel excavators operated by RWE Power in Germany are the largest mobile land machines on the planet. Fleet no. 288, located at Garzweiler open pit coal mine, is capable of excavating 314,000 cubic yards of material each day, and weighs a colossal 14,150 tons. This machine was built by Krupp of Germany and went to work in 1978 after a 4-year design and erection period. Note the truck and maintenance worker to the right of the machine.

After excavation by the wheel, material drops onto a conveyor which delivers it through the center of the machine onto a transfer conveyor. The excavator is carried on three 4-crawler assemblies (12 crawlers), two in front and one behind, giving the machine a stable 3-point support. The machine illustrated is fleet no. 285 jointly built by Krupp and O&K. It has a daily output of 260,000 cubic yards.

The mobile transfer hopper receives material from the bucket wheel via a transfer conveyor. The crawler-mounted hopper has its own operator who controls the swing of the discharge conveyor and the movement of the hopper. The hopper discharges onto a long conveyor running the full length of the face, one of a series conveying the material around the end of the open pit, and onto the reclaim side of the pit where it feeds a crawler-mounted spreader.

There's a lot of machinery involved in the mobile spreaders located on the reclaim side of the German open pit mines. Material is picked up from the main conveyor by a "tripper" at right. Then the material passes via a crawler-mounted transfer bridge to the spreader. Each spreader is dedicated to one bucket wheel excavator, and thus has the same production capacity as its respective "wheel." Built by Demag, this spreader has an operating weight of 4,942 tons, and swings a 328-foot boom.

Here is bucket wheel excavator fleet no. 284 built by Krupp. With somewhat less capacity than the giant machines, it still tallies a respectable 8,577 tons in weight, and handles 144,000 cubic yards per day—31 electric motors, totaling 12,370 horsepower, drive the main motions. The wheel dozer, dwarfed by the 58-foot diameter wheel, is essential to keep the working area clean and level. The wheel operator has a front-row seat in his cab located near the action at the end of the wheel boom.

The digging end of bucket wheel excavator RWE fleet no. 262, another rated at 144,000 cubic yards per day and located at Garzweiler mine. The 58-foot diameter digging wheel carries 10 buckets, each holding 4.7 cubic yards, and swings from side to side picking up all material within reach. Estimated operating weight of this machine is 8,422 tons. Conveyors taking material away to the dump area travel up to 17 miles per hour.

Bucket wheel excavator no. 293 is currently the world's largest machine moving on land. Built by MAN/Takraf, and starting work in 1995, it is the largest of eight bucket wheels operating at RWE Power's Hambach mine. The machine is capable of excavating 314,000 cubic yards of material each day, and boasts an estimated weight of 15,626 tons. The 70-foot diameter digging wheel, so often dwarfed by the overall size of the excavator, is shown here in comparison with mine personnel.